Do Beekeeping

The Secret to Happy Honeybees

Orren Fox

Published by
The Do Book Company 2015
Works in Progress Publishing Ltd
thedobook.co

A CIP catalogue record for this book
is available from the British Library

PB ISBN 978-1-907974-20-5

3 5 7 9 10 8 6 4 2

To find out more about our company,
books and authors, please visit
thedobook.co
or follow us on Twitter **@dobookco**

5% of our proceeds from the
sale of this book is given to
The Do Lectures to help it achieve
its aim of making positive change
thedolectures.com

The author will be donating his
proceeds from the sale of this
book to **BlinkNow.org**

Cover designed by James Victore
Book designed and set by Ratiotype

Printed and bound by Livonia Print
on Munken, an FSC-certified paper

'There are certain pursuits which,
if not wholly poetic and true,
do at least suggest a nobler and finer
relation to nature than we know.
The keeping of bees, for instance.'

—

Henry David Thoreau

For the beekeepers
at Kopila Valley School
and to Duke

Contents

Introduction

I opened the car door, walked to the back and opened the trunk, out of which I grabbed my bee suit, gloves and toolbox. After shutting the trunk, I began to make my way towards the hives. It was a warm, pleasant, late August afternoon. I glanced around as I walked, enjoying the little grove of trees where my bees live. It was rather humid and very buggy, so I put down my tools and stepped into my suit. I lazily tossed my veil onto my head to keep the bugs away, but didn't zip my suit because of the heat. On my feet, I was wearing a pair of flip-flops – a risky yet conscious decision. It was rare for me to get stung on my feet, so with this in mind, I opted for the ease of flip-flops rather than sneakers for today.

As I approached the hives, I noticed that the bees were incredibly active – something I had come to expect on a hot summer's day. Hundreds of them were coming and going from each hive every second. The apparent chaos at the front of each hive belied the fact that each and every bee knew exactly where it was going and what it was doing. Being a human, I found it hard to fully appreciate the organisation at work in front of me.

After making a quick observation, I zipped my suit, grabbed the necessary tools and headed over to the hives. As I walked nearer, their buzzing got louder. I watched as bees came and went, the sun illuminating their backs. As I prised the lid off the first hive, countless bees surged out and came crashing into my veil. It's an odd feeling to have hundreds of little insects greet me with such enthusiasm but one I have now become accustomed to.

The aromas of this first hive were both sweet and quite savoury. It's a smell that can only be found in a freshly opened hive, nowhere else. Once the delicious aroma faded, I started to check up on things, pulling out a few frames in order to get a sense of how the hive was doing. As I worked, I was relieved to see that this one seemed to be coming on quite well. The queen had been laying many eggs, the baby bees – known as brood – were abundant, and there was a vast amount of honey and pollen. And the hive was not overly aggressive, which came as a relief.

Of all the experiences I've had with bees, it's ones like these that give me most pleasure. Although the primary goal of visiting my bees is to observe the health of the hive, my underlying aim is merely to enjoy the moment. Beekeeping is unlike any other activity. While it can have its downsides, overall it is an extremely positive experience. It's hard to replicate the feeling of bottling your first jar of honey, or the moment when you realise your hive has survived the winter. Both are extremely gratifying and go some way to explain why I keep bees. Throughout the rest of this book, I hope you will learn more about these fascinating creatures and find your own reasons as to why and how you too should keep bees.

1
Bee Basics

'Concerning the generation of animals akin to them, as hornets and wasps, the facts in all cases are similar to a certain extent, but are devoid of the extraordinary features which characterise bees; this we should expect, for they have nothing divine about them as the bees have.'

Aristotle

The humble honeybee has played a significant role in human history. Depictions of honeybees can be found in cave art created by some of the earliest civilisations. Although our ancestors were not keeping bees at this time, they were still benefiting from one of the bees' many services: the creation of honey.

Beekeeping didn't become commercially viable – i.e. where the hive and bees no longer had to be destroyed to harvest the honey – until the 19th century and the invention of the movable comb hive, but humans had kept their own bees long before that. Some of the earliest indicators come from the tombs of ancient Egypt. Honey has been discovered inside these tombs and, perhaps even more impressive, the honey had not degraded! This is one of the amazing qualities about honey – it doesn't go bad. In both a literal and figurative sense, honey is timeless.

Over time, the honeybee's role in society has changed and evolved. Beekeeping is no longer mainly a hobby: it has, like so much agriculture, become industrialised. For the purposes of this book, we'll be keeping things on a small scale.

Hive Hierarchy

A hive of honeybees is a complex unit. It consists mainly of worker bees, all of which are female. There is also a group of drones, the males in the hive, vastly outnumbered by the female workers. Finally, there is the one and only queen, the mother of the hive.

One single hive of honeybees can have as many as 100,000 bees or as few as 15,000 – of which only 500 to 700 might be drones. The reason why there are so few drones is simple: they do not work. The drones' only task is to mate with the queen, and that only happens once in her life. After they have done this, the drones do nothing other than eat copious amounts of honey and pollen.

As the worker bees go through their life cycle of approximately 35 days, their role changes. When a worker is born, her first task is to clean out the cell where she was incubated. After this, she remains in the hive as a nurse bee – feeding the brood – or as one of the queen's many faithful followers. When the worker bee grows older, she becomes a forager. Every day she will fly out of the hive in search of nectar and pollen.

The queen is rather easy to find in a hive due to her size. She is significantly larger than the workers and drones, nearly twice the size. Still, some beekeepers often put a spot of paint on the queen so she can be easily identified. The drones are bigger than the workers, but not as majestic in size and stature as the queen. Regardless, all bees are important, and each has a crucial role in the success of the hive.

The Life Cycle of a Bee

The life cycle of a worker bee is rather simple. It begins with the queen, who lays an egg at the bottom of a wax cell in the hive. The egg turns into a larva within three days, and this larva is fed by her older sisters, the nurse bees. Before long she will turn into a pupa, where she grows significantly and finally begins to look like a bee.

The males take slightly longer to exit their cell and, because of their size, they require larger cells. When you look inside your hives, it's easy to see the drone cells because they jut out from the rest.

Queens are by far the most interesting of the bees and their cells, known as supersedure cells, are the largest. In fact, they are so large that they cannot fit into the comb itself and the bees build a separate queen cell that hangs off the bottom of a frame in the hive. A queen bee begins life as a normal worker bee, but instead of being fed honey when she is young, she is fed royal jelly, a secretion used in the nutrition of larvae. This special diet is the only thing that changes a normal worker bee into a majestic queen bee.

Workers live for the smallest amount of time, just over a month. Drones are slightly luckier and usually live for a few months unless they mate with the queen, in which case they die immediately. A queen, however, can live for a few years until she dies or is replaced. These are the basics of a beehive, at the same time a complex system and a simple community meshed into one.

An interesting thing to note is that each bee has a unique job depending on its sex and its age, from the queen to the guard bees.

There are three types of adult bees – the queen, the drones, and all the workers. The worker community has unique physical attributes such as pollen baskets, wax

Fig 1. Types of bee

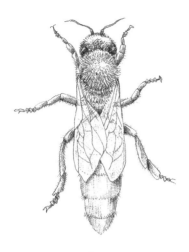

Queen

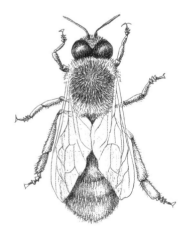

Drone

Worker

glands, scent glands, and brood food glands, all of which help them do the work of the hive. As the bees' glands develop, their jobs change.

The worker bee starts out as a housekeeper, then after a few days she becomes a nurse bee. This is a very important role. She cares for the egg all the way to it becoming a larva. Her primary job is to create something called 'bee bread' or food for larvae as they rapidly grow. In addition, she will clean and feed the queen, clean the cells, manage the incoming nectar, and build out the wax comb. It is also the responsibility of the nurse bee to maintain the temperature of the hive at 95°F (35°C). Another big job! When the hive is too hot, the nurse bees work together to create a little air movement with their wings, and in the colder months they huddle together to create a mass of heat.

As the worker bees mature, some become field bees and forage for water, nectar, pollen, and propolis. Others become guard bees whose only job is to protect the hive. The guard bee will stand on her back four legs with the front legs raised and inspect every bee entering the hive. Each hive has a distinct odour and this smell is how the guard bee can tell if the bee is from her hive. Guard bees will sting any intruders including skunks, racoons or beekeepers!

The only job of the queen is to lay the eggs. The queen normally lays one egg per cell. The egg sits at the bottom of the cell and looks like a grain of rice.

Bee Anatomy

Physically, although bees are pretty complex, they are not unlike many other insects. So rather than write an anatomy lesson on the body of a bee, which would likely lull you to sleep, instead I'll give you a quick description of the parts you should know about.

Fig 2. Anatomy of a bee

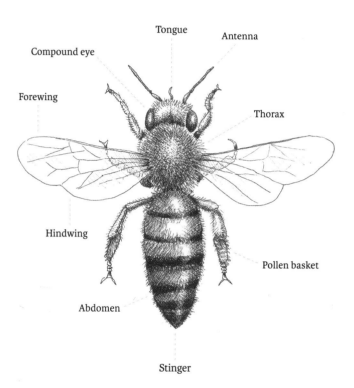

Tongue

Antenna

Compound eye

Forewing

Thorax

Hindwing

Pollen basket

Abdomen

Stinger

First and foremost is the stinger. Not surprisingly, the stinger is located at the very end of the abdomen. Unlike wasps, honeybees can sting you only once. When they sting you and fly off, a venom sac is ripped away from the back end of the abdomen, causing the bee to die. So when a honeybee chooses to sting you, she has decided that you are a serious threat to the hive. The venom sac is something you should look for when you are stung, because if you don't remove it with the stinger, it will continue to pump venom into the area and make a small problem far worse! So do try to remove it immediately.

On a happier note, located on the legs of honeybees are pollen baskets, used for carrying pollen to the hive. One of the greatest sights to witness as a beekeeper is a forager returning to the hive laden down with full pollen baskets. It's easy to spot: just watch the front of your hive and you will see bees landing who have a yellowy-orange powder attached to the backs of their legs. Of course, there are other important parts to bees – not least their sucking tongue, antennae and compound eyes! – but venom sacks and pollen baskets are the two that you should know about.

Hive Communication

One of the lesser-known things bees do is the waggle dance. The waggle dance is the bees' method of communicating, which involves, yes, dancing! When a bee returns to the hive after collecting pollen and nectar, she will do a waggle dance. This dance seems rather simple, a series of slight movements throughout the bee's body, but it contains a great deal of valuable information. The dance communicates the location of a great source of pollen or nectar which means other bees do not waste valuable time looking elsewhere. It serves as a language, allowing the bees to identify specific locations. For example, if a forager finds a bush that has just begun flowering and is full of pollen and nectar, she will disclose this location when she returns to the hive so that her sisters can find the bush and harvest from it.

As you can see, while they are tiny, bees are very complicated creatures and quite fascinating, which is why I recommend you acquire some of your own.

Summary: Bee Basics

— Worker bees are female, small and live for about a month. They are responsible for feeding larvae, hive management and foraging.

— Drones are the males, slightly larger than the workers, and can live for a few months. Their sole job is to mate with the queen.

— There is only one queen in a hive. She is the largest of all the bees.

— Worker bees communicate the location of ample sources of nectar and pollen by doing a waggle dance!

2
**Setting Up
Your Hive**

> 'He is not worthy of the honeycomb,
> that shuns the hives
> because the bees have stings.'
>
> —
>
> *Locrine*, Anon.

Where you decide to locate your hives can eventually determine whether your bees survive or not, making it one of the most important aspects of getting set up. In choosing a location, you need to keep a few requirements in mind.

The first is ventilation. It's extremely important that your hives have some solid air circulation, but also a bit of wind protection. In other words, you want to put your hives up against some trees, not in a huge open exposed area. This is crucial because if the weather becomes too violent and windy, your hives could topple over; placing them against a collection of trees is ideal since the trees serve as a partial windbreaker.

The other element to consider is sunlight. Not only do trees offer protection from the wind, but they also provide a bit of shade. However, you don't want your hives to be completely in shade, as they need the sun, especially in the winter. So you really have to use your best judgment to find a place where the bees will be warmed in the winter but not baked in the summer. It sounds contradictory and difficult, but I can assure you that finding the perfect spot isn't as hard as it seems.

Another thing to be conscious of when choosing a location for your hives is access to water, especially in the summer because bees use the water to help keep the hives cool. For example, my hives are located near some wetland, which provides the bees with a great water source. If you're setting up in an urban area, be mindful that bees are attracted to chlorine and will gladly head to an open-air swimming pool, looking for water. This can create havoc as the bees get stuck in the surface tension of the water and struggle to get out. Not something any swimmer wants to come across! In general, it isn't very difficult to locate your hives near a source of water. Something as simple as a bird bath can work as long as it's always kept filled. On the other hand, you shouldn't position your hives in a particularly moist place. Too much moisture can cause the demise of your hives, especially if you live in a cold area.

Cost

Bees are certainly a bit of a monetary commitment. However, most of your spending will occur up front when you buy the hives themselves and your tools. Once you have your hives set up, the running costs can be minimal, so don't let this expense discourage you from getting bees. In fact, if you choose to sell some of your honey, you can earn some money back. Yes, it might take a few years, but it can happen. If you lose a hive at any point, buying a new one will set you back again, but you get a new hive! Additionally, you will want to replace some of your equipment every few years.

Time

While a beehive doesn't need very much of your time, it can easily occupy a lot of it. When I visit my hive, it isn't usually for a particular reason other than that bees are extremely interesting to be around.

That said, immediately after installing a hive, it's important to give it a lot of attention. During the first few weeks, the bees are still getting settled in, and are establishing themselves – something that should be closely observed. During these first few weeks, you will learn more about your hives than at any other point. This settling-in period is crucial to hive health, so if you observe anything out of the ordinary, make sure to address it as quickly as possible. If you don't act quickly, it could have a negative effect on the whole hive and you may have to start again.

As time goes on, you will become accustomed to what is 'normal', so it will be obvious when you spot something abnormal. What you see may not be something problematic, and these moments of discovery are great ways to learn about your bees. Remember this: no matter how much you know about bees, there is always more to discover, which is why it is crucial that you approach your hives with an open mind, ready to learn.

Neighbours

Another very important thing to consider before you get your first hive is your neighbours. It certainly helps to be on good terms with those living close by, not least so that when you tell them that 50,000 newcomers are moving in soon, they will be less likely to freak out. Additionally, the promise of a fresh jar of honey can't hurt either. The main thing you want to establish is if any of your neighbours are allergic to bee stings, as that could present a problem. When I told my neighbours I was getting bees, they were ecstatic, genuinely excited about the prospect of an annual jar of honey as well as the potential experience of suiting up and visiting the hives themselves.

Stings

As we're talking about setting up, now is as good a time as any to bring up the subject of stings. Unfortunately, it is really hard to avoid getting stung if you acquire bees. Yet while they may seem intimidating, stings are honestly not that bad. Yes, it is painful to be stung, but it's merely part of the job. You can, however, take precautions to limit the amount of times it happens. For example, most beekeepers wear shoes. I choose not to wear sneakers during the summer when I visit the hives out of sheer laziness. This inevitably leads to me occasionally getting stung on my feet, which happens to not be as painful as other places.

It is possible to make yourself sting-proof. The simplest way to avoid stings is to always wear your bee suit when you visit your hives and make sure that your suit is completely zipped up and your gloves pulled tight around your arms. These two little details are quite effective if you're trying to stay sting-free. If you wear socks and sneakers, tuck the bottom of your bee suit into your socks in order to avoid having bees fly up your trousers (fortunately this has never happened to me). Wearing long sleeves and long trousers or leggings under your bee suit serves as yet another layer against the stinger, but doing so does have the unfortunate side effect of making you quite hot.

Unless you are allergic to bee stings, they are honestly not as bad as you might think. If you remember to pull out the stinger immediately after getting stung, swelling will be kept to a minimum. The surprise is often the worst part. They can get you while you are innocently meandering back to your car for the drive home. They can even seek you out while you are in your car! However, do not be intimidated by bee stings. Your bees are not out to get you – they would much rather let you mind your own business than sting you.

Summary: Setting Up Your Hive

— Position your hives where they are not exposed to strong wind and sun.
— Ensure your bees have access to water, especially in the summer months.
— Be prepared for an up-front cost.
— Remember, selling your delicious honey can offset this!
— Be mindful of and courteous towards neighbours.
— Take measures to minimise the risk of being stung.

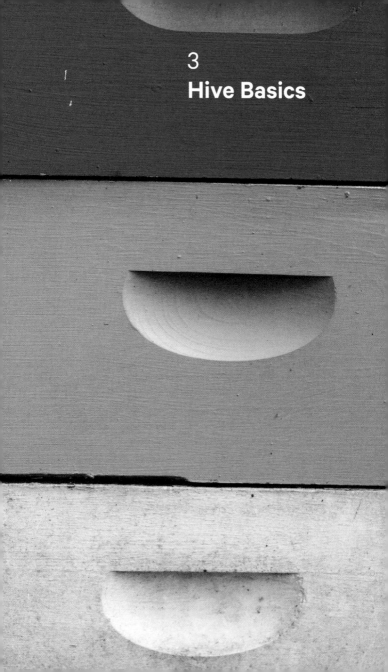

'Bees do have a smell, you know,
and if they don't they should,
for their feet are dusted with spices
from a million flowers.'

Ray Bradbury

**The structure of a beehive is actually remarkably
simple. When you approach a conventional beehive,
it looks like a few white boxes stacked on top of each
other. This isn't terribly far from the reality.**

Supers

Each of these mysterious white boxes is called a 'super'.
The bees live in the larger ones, called deep supers. The
larger area provides the perfect amount of room for the
bees to conduct their business. If you visit a hive during the
summer, you will likely see some smaller white boxes sitting
on top of the larger ones. These are honey supers, not for
the bees to live in. If the honey supers were the same size
as the deep supers, they would weigh a lot when filled with
honey. Even being the size they are, they are tough to carry
when the bees have filled each cell with their golden nectar.

The cell is the foundation of the beehive. These are
hexagonal in shape so that many can fit together side by
side – and are perfectly designed for their purpose. A cell
is both where the brood is stored and where the honey and
pollen are collected, but not in the same cell. Each cell has
its own unique purpose.

Inside each super are ten frames. These frames vary in size, just as the supers do. Obviously, there are shallower frames for narrower supers and deeper frames for wider supers. A frame is made of four pieces of wood in the shape of a rectangle. Inside each rectangle is the wax base. Each of these consists of a very thin sheet of wax with the honeycomb pattern imprinted upon it. This sheet of wax is where the bees begin constructing their cells. While I prefer to use wax sheets as the base of my frames, some people use slightly thicker sheets of plastic that have a thin coating of wax on top. This plastic allows the frames to last longer than the pure wax ones. I tend to use pure wax sheets as I have chosen to use natural practices. However, I have found they don't last beyond one harvest, while the plastic sheets last for many seasons. Both bases are great options and work very well, so it's really up to you to decide which one you want to use.

The Hive Base

For the base of your hive, you will need three pieces. At the very bottom, you absolutely must have a sturdy, reliable stand to put your hives on. I have one at the bottom of each of my hives and they are fantastic. I was fortunate; my father is a furniture maker and he built some very deluxe bases for me. However, something as simple as an old wooden pallet works just as well, provided it's in good condition and is strong. Keeping your hive off the ground is very important. It helps prevent moisture accumulation and keeps other animals out. Without a proper base, water could flow freely inside, mice could stake out a warm house, and many other problems could arise that a basic stand would prevent.

The second thing you will need for the bottom of your hive is a landing board. The landing board is a rectangle, but one of the sides is slanted inward, giving the bees a place where they can land safely. When a bee returns from a foraging trip, she is packed full of supplies, understandably making it a bit tougher to fly. The landing board allows her to land safely, then walk up into the hive. There is one thing to be wary of when dealing with your landing board. When this piece is placed under your hive, you will notice that it creates a small gap, about 6 inches (15cm), between the bottom of your hive and the hive stand. This space is where I have had the most problems with mice. Imagine this: you are a field mouse looking for a home in the dead of winter when you stumble across a beehive. The hive is nice and warm, and it provides protection from the elements. If you were to enter the hive, though, the bees would sting you many times, and possibly kill you, so you keep looking around. This is when you find a nice area under the hive – somewhere equally well sheltered and warm. Why would you not want to stay there? This is a great reason to check under your hive because if you don't, it could turn into a mouse hotel.

The final part of the exterior of your hive is the roof. Most people use a completely flat roof on top of their hives, but I have transitioned to a triangular or 'garden roof'. I decided to do this because a garden roof allows water and snow to slide off easily. They are also made of copper and look great.

Other Hive Components

Other parts of your hive include an entrance reducer, possibly a queen excluder, and definitely an inner cover.

The entrance reducer is a very simple piece of wood that you put in the front of your hive. This can be cut in various different ways, but mainly serves to keep draughts

Fig 3. Structure of a beehive

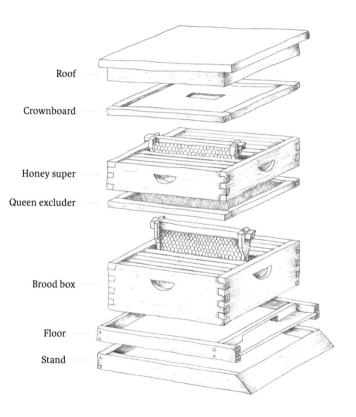

Roof

Crownboard

Honey super

Queen excluder

Brood box

Floor

Stand

out in the winter or protect a weak hive from any invaders. Entrance reducers are not crucial to your hive, but can be very useful for filling in spaces or used as a tool in other circumstances. I highly recommend having a couple extra in your toolbox at all times.

A queen excluder is a piece of metal grating that you place in between your deep supers and honey supers. Its purpose is to prevent the queen from laying eggs in the honey supers. The pieces of the grating are large enough to let the worker bees through, but small enough so that the queen cannot squeeze through – hence the name. Some people, myself included, choose to use these. Others don't, because they are said to slow down your bees. If you want to take the risk of having your queen lay in your honey supers, feel free to not use one, otherwise opt for the safe route. I use a queen excluder and offset the potential loss by starting the honey-gathering sooner rather than later. This decision is like many others you will have to make as a beekeeper. For the most part, there is no correct or incorrect way to do things: it is entirely up to you and you will discover what works best for your hives over time.

Last but not least is the inner cover, also known as the 'crown board'. This is an essential part of a beehive. Located between the top super and the roof, it serves as an entrance for the bees, allows correct airflow throughout the hive and ensures the correct amount of bee space between the body and the roof of the hive. The inner cover has many different uses. During the summer, it is common to see the roof of the hive propped up at one end to allow the bees easy access to a small entrance at the front of the inner cover. This is not necessary, but the bees love it and use it all the time when they can.

Initially, your hive will appear to have a correct and incorrect setup but, as you will slowly learn, each hive is like a home. It is individual and has its own quirks which can be adapted and improved by you, the beekeeper, over time.

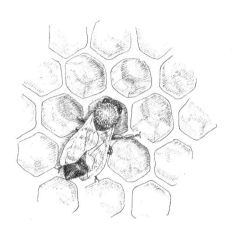

Summary: Hive Basics

— Hives are made up of box-like 'supers'. Bees live in the deep supers.
— Honey is stored in the smaller 'honey supers' placed on top in the summer months.
— Inside each super are ten rectangular frames with wax bases.
— Bees construct their cells and store honey and pollen inside. Within separate cells, eggs are laid, and larvae and brood develop.

4

Suited and Booted: Your Beekeeping Kit

'Agriculture ... is our wisest pursuit,
because it will in the end
contribute most to real wealth,
good morals, and happiness.'

Thomas Jefferson

Every time you visit your bees, you'll need to take a few basic items with you to make your life a lot easier. The first and most important of these is your suit. Putting on a bee suit for the first time is hugely exciting. It looks similar to a hazmat suit and, when showing friends and family photos from the beehives, I have been compared to an alien, a bomb disposal expert, and even the Pillsbury Doughboy.

The Bee Suit

If you were ever scared about being stung while tending to your bees, the suit will immediately rid you of those fears. For me, the comfort it provides is half psychological and half physical. Just having the suit cloaked over you gives you the confidence to dive right into a hive with no reservations. My only advice would be that before you suit up and get rolling with your hives, take a moment to familiarise yourself with it. I learnt this the hard way as you'll discover later!

Gloves

Another obvious part of the attire you will need while caring for bees is your gloves. There is little variation with the gloves you can buy, but they all do the trick. Over time they accumulate propolis (a sticky mixture that bees collect from tree sap and other sources used to seal gaps and holes in the hive, and help to protect the bees from varying weather conditions), wax and honey, which makes them pretty appealing to the bees. As my gloves have been gradually covered with these materials, my bees have become more and more attracted to them. Now, after any hive inspection, I look down at my gloves and they will be absolutely covered with bees.

I would recommend buying a pair that are slightly too big, merely for some added comfort – or if you're a young beekeeper who isn't done growing yet! I am already on my third pair of gloves because of my crazy growth spurt over the past two years. However, for you older beekeepers this shouldn't be a problem, so get a pair of gloves that fit and stick to them.

Smoker and Fuel

One of the crucial tools you'll take each time you visit your hives is your smoker. I choose not to use mine very often, but it is a very effective tool when trying to manage a particularly angry hive, so I always have it with me. Plus I love carrying it! It is probably one of the most iconic and photographed pieces of beekeeping equipment.

In order to get your smoker going, you will need some fuel. The key to good smoker fuel is to use something that will produce tons of smoke and not actually burn very well. If your smoker is filled with a blazing fire and no

smoke, it won't do you much good. I'd recommend pieces of a burlap sack, pine needles, sticks, newspaper and anything else you can find that will get really smoky.

Frame Hangers, Bee Brushes and Hive Tools

Other items in my toolbox include a frame hanger, something I didn't have during my first year as a beekeeper, but a tool I can't imagine working without now. It is a very simple metal object that you place on the side of your hive to act as an extra hand. Before acquiring this item, I would lean my frames against the bottom of the hive. Occasionally they would tip over, or I'd accidentally kick them; as you can imagine, this would really upset the bees. The other perk of having a frame hanger is that it allows you to inspect your frames far better than you could without one. When you've finished an inspection, you can hang the frame up on this holder and move on to the next one. While it is not a necessity, a frame hanger can be extremely helpful.

Two items that are essential to a beekeeper's toolkit are a bee brush and a hive tool. They are both very simple, yet highly effective.

The hive tool is a piece of metal with a hook on one end and a chisel-like piece on the other. It is used to open up the hive and move frames around with as much ease as possible. The hook allows you to pick up frames out of your hive in the smoothest, least destructive way possible. The chisel is used to remove any propolis that might be standing in your way. This tool more than any other is something that you cannot do without.

The bee brush is an item you will want to have on you at all times. It is exactly what it sounds like, a very simple, gentle brush used to move bees out of areas where you don't want them. I mainly use this when I'm putting a hive back together after inspecting it. As I'm stacking the supers back on top of each other, I don't want to be crushing my bees, so I use the bee brush to move them off the edges of the hive back to safety. These two tools are my absolute favourites. I'd advise you to find a set of tools which work for you and stick with them; you won't be disappointed.

Keeping a Journal

Keeping a journal is never a bad idea, especially when it comes to beekeeping. I will be the first to admit, I keep a very loose journal and could definitely write more. I just write down some of the more important observations so that I don't forget them. If you observe something out of the ordinary or that you simply find interesting, try to get into the habit of writing it down. You may need to refer back to it. If you have more than two hives, I would highly recommend keeping a journal.

Summary: Suited and Booted

Basic tools and equipment you will need include:
— A suit
— Gloves
— Smoker
— Smoker fuel
— Frame hangers
— Bee brush
— Hive tool
— Journal (optional but useful)

5
**Installing
Your Bees**

'The keeping of bees is like
the direction of sunbeams.'

———

Henry David Thoreau

If there is one thing I am certain about when it comes
to bees it is this: installing your first package of bees is
unlike any experience you have ever had, or ever will
have. After I acquired my first three packages of bees,
I put them in the back of the car and drove home.
When we arrived, I carefully carried the packages into
our house. I watched the buzzing clusters of bees inside
for a few minutes and listened to the soothing sound
coming from each package. I couldn't part ways with
such a new and remarkable sound, so I took one of the
packages up to my bedroom and placed it against the
wall. The bees continued buzzing as I drifted off to sleep.

I was fortunate enough to live near two very experienced
beekeepers who were glad to help me out the next day.
Without them, the process of getting the bees out of
the packages and into the hive would have gone very
differently! I would highly recommend trying to find other
beekeepers in your area to help you get started, otherwise
it could be rather stressful. I would never have had the
confidence to get the bees out of their transportation box
and into the hive on my own. Instead of trying to coax the
bees out, I quickly learned that you have to force them out.

This is one of many, many things my beekeeping mentors have taught me.

Package or Nuc?

When purchasing your first bees, you have two choices: you can order either a 'package' or a 'nuc' (or, if you're feeling brave, you can capture a swarm! This method is of course free, but you never know when or where it will happen).

The difference between a package and a nuc is this: a nuc of bees comes already established in a little hive with frames of brood, a queen, and workers. It usually comes from a local beekeeper and they share with you the nucleus of one of their hives.

A package, on the other hand, includes a queen and worker bees who are simply waiting within a smaller transportation box – like a small shoebox with a screen – that is often shipped in the mail from a 'grower'. This means you have to get them out of their transportation box and into the hive yourself.

I prefer packages simply because I've had the most success with them. I have tried nucs, but sadly they have all died rather quickly, so I reverted to packages. Nucs are appealing because they are more established. However, they can also be a bit more expensive. Some beekeepers will only purchase packages and others will only purchase nucs. The choice is yours: like everything with beekeeping, find what works for you and stick to it.

Installing Your Bees

Perhaps the best way of describing the process of installing bees is to go back to my first experience.

I was excited as I strolled along through the tall grass. I held a toolbox in my left hand and a smoker in my right. It was midday, and the sun was beaming warm, bright light down upon the grove where my hives sat. I was way ahead of the friends and family who had come along to witness my hive initiation.

When I arrived at the empty hives, my heart was racing; I was so excited to get them going that I was almost shaking. I set down the toolbox, opened it up and picked out a box labelled 'Smoker Fuel'. I selected some suitable

fuel chunks and placed them into the smoker. Then I grabbed the lighter out of the side of the toolbox and began lighting the fuel. I gave the smoker a few pumps to force a bit of oxygen onto the growing fire. As I did this, a few soft plumes of smoke rolled out of the top of the smoker. As the summer sun hit those little clouds, I knew that I was going to enjoy this.

Roughly fifteen minutes later, I was prying the top off a box of bees, ready to introduce them to their new home. I carefully pulled out a can of sugar water then slid the cover back on the box. Inside were roughly 30 to 50,000 bees. Once the can was outside the box, I poked a few large holes into the top and poured the sugar water into a spray bottle. I was then instructed by two experienced beekeepers – who had offered to help me install my hives – to spray the mass of bees inside the box with this sugar water. I did as I was told and sprayed the bees down.

Next, we banged the side of the box, forcing most of the bees to fall to one side. Then we lifted the box and poured the bees into the hive. I'm not exaggerating when I say pour. Since the bees were covered in sugar water, they were very sticky and stuck to each other. When we tipped the box, a glob of bees rolled out into the hive. It was one of the weirdest experiences I have ever had with bees. At first I thought that pouring them out would injure them, but they were completely fine. Within a few minutes, they had shaken off the sugar water and were buzzing around as if nothing had happened. It was a very effective way of getting the bees into the hive, but also very bizarre.

After we finished the first hive, we began to do the same to the next. There were thousands of bees buzzing around, and it was exhilarating. I was covered in bees! The ones we were installing were babies, and they love to cling.

All was going well when, midway through the second hive, I felt something on my stomach. I ignored it for a minute or two, then noticed it again. It felt like someone was tickling my tummy. I was tempted to just reach down and scratch it, but my instincts told me that was a bad idea. Instead, I told my companion that I wanted to check something, and brushed myself off. Once I had removed most of the bees from my suit, I slowly unzipped the front. When I lifted my shirt, I was greeted by about eight little bees looking up at me. I brushed them off and zipped my suit back up. Once I was all set, I wondered how they had got inside. Frantically, I looked over my suit for any holes.

Then I found them. For some odd reason, the bee suit had two sets of pockets. The first was a normal pocket, attached to the suit itself. The second, however, was merely a hole in the side of the suit in order to allow you access to the pockets of your own clothes. Yes, you read this correctly, the bee suit had holes in it, on purpose! I understand what the designers were thinking, but hello! A suit that is supposed to keep nosy little insects from stinging your body has two large holes in it? That doesn't sound very good to me. Luckily, I stayed calm when I realised what was going on, which is probably why I didn't get stung. If you get riled up around your bees, you are far more likely to be stung than if you stay calm.

This piece of advice is something worth applying to getting stung. When you are stung, the bee that stings you releases a pheromone that signals 'danger' to the rest of the bees. If you freak out, that only compounds the problem, making you even more likely to get stung. Although it will be tough during the first few times you visit your hives, always try to remain calm and both you and your bees will be happy.

Beekeeping Associations

Fortunately, I live near a beekeeping association whose members kindly informed me where I could purchase my first package of bees. I highly recommend searching for your local beekeeping association; as I have found out, it can be a great source of information and advice.

Before acquiring my first hives, I attended weekly meetings at my beekeeping association. I used to call this going to 'bee school'. At bee school I heard many lectures, listened to various speakers and learned so much about beekeeping. There were speakers who focused on specific areas a novice beekeeper is likely to encounter, such as foulbrood (of which more later), and others who gave more general tips and advice. The sessions were extensive, very thorough and extremely informative, so I would definitely recommend finding your local beekeeping association and learning as much as possible from the most knowledgeable beekeepers in your area.

Summary: Installing Your Bees

— You can purchase a 'package' or a 'nuc' of bees to get you started.
— Installation can be tricky at first so try and find an experienced beekeeper to help.
— Find a local beekeeping association. They are a great source of help, advice and information.

6
**Opening
Your Hive**

> 'The bee ... gathers its materials from the
> flowers of the garden and of the field,
> but transforms and digests it
> by a power of its own.'
>
> -----
>
> Leonardo da Vinci

After my first hive installation, I found myself worrying constantly about the hives. Had I installed them correctly? Did they have enough water? What if I had accidentally killed the queen? (In reality this last situation would have made things quite difficult.)

Rest assured that, for the most part, your hives will be doing perfectly well left to their own devices. If you begin opening them too soon and too frequently, you could do more harm than good. You need to tread a fine line between inspecting hives too often and not checking them enough.

During the first few weeks, you will want to keep a watchful eye on your hives. A check can be as simple as watching the front of each hive and ensuring that there is activity. If there are hardly any bees coming and going from the front of a hive, you will want to open it up and see what's going on. And we'll go through how to do just that in this chapter.

After your hive is settled, a check every two or three weeks is adequate; you will be able to track the progress of the hive while not bothering the bees and upsetting their home too much. Obviously, this time period can be reduced if you find something out of sorts within a hive (of which more in Chapter 9). Similarly, if you are unable to visit your bees for a little while, they will probably be just fine. In cases of extreme weather, you might have reason to be concerned and may need to ask a fellow beekeeper to check on your behalf, but otherwise your bees will take care of themselves perfectly well.

When to Open Your Hive

Bees are busy creatures – they work nearly all day. From roughly 10 a.m. to 6 p.m., the hive is vacated by thousands of worker bees heading out on their daily commute. Between these times is the perfect opportunity to check on your hives; there will simply be fewer bees around so you will have much better access. If you open a hive during the morning or evening rush hour, you will find yourself with a few extra thousand bees who would like you to leave them and their hive alone. So it's best to avoid these times. If this is the only time that you can visit, try to minimise your presence by being very gentle yet working quickly – this can be easier said than done.

Opening Your Hive

The process of opening a hive for inspection is remarkably simple. Things become more complex once the hive is opened up, as there are many things to look out for while attempting to take as many mental notes as possible. The ideal scenario would involve two beekeepers, one to observe what's going on within a hive and the other to take notes on what the observer sees. If this is not possible, take photos and simply observe what is occurring within each hive to be transcribed into your journal later.

I tend to do this for all of my hives and, so far, they have done quite well. Additionally, by simply taking mental notes and a few pictures, I have learned so much about my hives because I was completely focused on mentally recording what I was seeing.

Opening a hive begins with the removal of the hive covers. This process involves no more than a hive tool and a little effort. The outer cover should come off with ease, but the inner cover requires some prising. Your bees will have a tendency to use propolis to secure the inner cover. On hot summer days, this waxy substance is easy to manipulate since it becomes quite gummy with the extra heat.

Use the hive tool to cut away any propolis that is preventing you from inspecting the frames. I have described the hive tool in a previous chapter, but it bears repeating that it is the most important tool you will use. This opening and inspecting process is where the hive tool really comes into its own. Without it, a beekeeper will have a very difficult time removing the frames and inspecting a hive.

After you have removed the outer covers, I would recommend placing them upside down on the ground, which allows them to double as a table. Whenever I open up a hive, I always lay my tools on top of the covers since they provide a really useful, clean, flat surface. Next, begin to remove the frames. As you lift each frame out, you will need somewhere to safely and conveniently put them while you're going deeper into the hive. For my first couple of years of beekeeping, I simply rested them against the hive stand. However, this was not very convenient and it slowed me down quite a bit, so now I use a frame hanger, which attaches to the side of the super. This simple piece of metal allows you to store your frames in a convenient spot while you are working within a hive.

Me, age 14!

Inspecting Your Hive

Once inside, I always begin by removing the frames at the outer edges first. By doing this, I provide extra space within the hive, which allows me to move the remaining frames around with ease. Additionally, by beginning with the outer frames, I immediately get a sense of how the hive is doing. If the bees are active in the outer frames, I know the hive is likely to be doing well. However, if there is very little activity in these frames, I begin to get worried.

In the outer frames, the bees will mainly store honey and pollen, and occasionally you will find brood. If this is what you see, the hive is healthy and doing well. The queen lays the vast majority of her eggs in the centre of each super, so if you find eggs or brood in the outer frames, it means the queen is laying really well. After the outer frames are removed, work your way to the centre. The egg of a queen bee looks remarkably like a grain of rice. At first, the eggs are difficult to see, but they become easier to spot with practice.

The centre of each frame is where the main activity is found. Within the middle four frames there will be vast amounts of honey, pollen and brood, not to mention an abundance of bees. While you might be tempted to see a frame and think there is no brood, look closely. Uncapped brood is white and looks eerily similar to maggots. However, don't be worried – these white little creatures will soon resemble a honeybee as we all know it, with black and yellow stripes and big wings. After the brood has matured enough, the nurse bees put a wax cover over each cell. This wax capping prevents viruses and bacteria from entering the cell.

Checking up on the amount of brood is one of the main reasons to open your hives every few weeks. During these checks you will not only look for brood but for honey and pollen as well. If you see honey and pollen everywhere yet no brood, your queen might have died; this is obviously a problem.

Without a properly laying queen, a hive can be lost within a matter of weeks. I'm always quietly relieved whenever I spot a queen crawling around within a hive. You have to be quite lucky to see this, and whenever you do it's a great moment. I also get excited when I find many freshly laid eggs all together in one neat bunch. This is the end result of much hard work from a productive and successful queen bee.

However, in due course every beekeeper will encounter supersedure cells hanging off one or more frames. A supersedure cell is what a hive will create when a new queen is being raised. They may want to replace an existing queen for several reasons: she isn't laying enough eggs, she is weak, or she is simply old. When the queens hatch, they will fight to the death and the strongest becomes the new queen for the hive.

When I first encountered supersedure cells within one of my hives, I was overly curious. I couldn't believe that an actual queen bee was hiding within what appeared to be a completely dormant cell. Naively, I decided to open one of the cells simply to observe what was occurring within it. As I uncapped the cell, I saw movement inside. Then, as I carefully removed more of the wax, the queen began to crawl out! Soon a brand-new queen bee was sitting in the palm of my hand. I couldn't believe it! I figured it must have been late in the 24-day gestation period, as she appeared to be fully formed and quite healthy. As she crawled around, I observed her every move. She truly was magnificent, large and powerful. Her body was long, built for laying eggs.

Once the astonishment of having such a creature crawling in my hand had passed, I began to worry that I had just ended a queen's life. Had my curiosity cost my hive a completely healthy queen? I decided not to dwell on this; instead I placed her back into the hive, hoping that in doing so I would allow her to survive. I was further reassured that the hive would be all right when I spotted a few other supersedure cells hanging off another frame, each containing a queen. In this one moment I learned far more about a queen bee than I could simply from reading about it in a book.

Summary: Opening Your Hive

— In general, check your hive every two weeks.
— Timing is everything. Avoid opening your hive during the morning or evening rush hour.
— Make sure you have all your tools and smoker to hand before you start.
— Begin with the outer frames and work your way inside.
— Closely observe everything, compare it to your last check, and note down any changes.
— Move slowly and gently!

7
**Feeding
Your Bees**

'Bees work for man, and yet they never bruise
Their Master's flower, but leave it having done,
As fair as ever and as fit to use;
So both the flower doth stay and honey run.'

———

George Herbert

Most of the time your bees will be able to feed themselves just fine. They are wonderfully self-sufficient and live off the honey they produce. However, sometimes – particularly in the winter months when honey stores could be running low, they might be a smaller hive or perhaps they weren't able to pack enough away – they could use a helping hand, just like the rest of us. I usually end up feeding the bees in the first year, but by the second year they are up and running.

Sugar Water

Obviously, humans cannot create honey, so sugar water is used as a substitute. You can easily make this syrup at home by combining the cheapest sugar available with hot water. Use a one-to-one ratio of sugar to water; I use one cup of water to one cup of sugar. If you are feeding during the warmer months of the year, you want to use a thinner syrup, which consists of roughly two parts of water to one part of sugar.

The actual process of making sugar water couldn't be easier. Simply heat up some water to boiling point, pour the sugar in, and stir it around until dissolved.

Then pour it in jars ready to feed to your bees. The most straightforward way of doing this is by putting some small holes in the lid of the jar and placing it on top of your inner cover with an empty super placed around it for protection from the elements. The holes shouldn't be too big or too small – you don't want the sugar water to flood the hive. I use the tip of a nail to create a hole of about the right size.

Although this simple syrup is a great way to supplement your bees' food if they are lacking honey, it isn't perfect, so proceed with caution. If you give it to your bees late into the autumn or even early winter, it could prove lethal for the hive. The bees must 'cure' the sugar water in order to keep it from fermenting, so if you don't give them enough time to cure it and they pack it away into their hive, it could make them ill. Additionally, if you give your bees the syrup when the weather is too cold, the water could condense in their hive and create all sorts of problems. It can cause the moisture levels to rise in a cold hive – as I was to discover during my first winter as a beekeeper.

I remember the day very clearly. It was a bright day in the middle of December. I had my jacket zipped up tightly around my neck and my trousers tucked into some warm Bean boots. While it was cold, the sun provided nice little pockets of warmth. The snow crunched under my boots as I trekked towards the hives. When I came to the clearing, I saw the hives completely buried under a layer of fresh, white snow. All I could see were the rooftops peeking through. The lower level of the hives were a few feet under.

I got to work as quickly as I could. As soon as I'd pushed the snow pile from the top of the first hive, I began to shovel out the entrance. When I had cleared most of the snow away, I put my ear next to the hive – nothing.

This confirmed my initial fear; the hive was lost. When I cracked open the top, a light dusting of snow swirled up into the air. As it cleared I could see thousands of bees, all lying completely still. It was unlike anything I had seen before. The hive that I had only ever seen bursting with life was now completely full of dead bees. Inside the hive appeared blue; it was cold and slightly damp. The snow that had buried the hive had raised the internal moisture to a fatal level, and now my first hive was gone.

At first, it made me very sad. I blamed myself entirely. How had I let something as powerful as a beehive lose its way in the cruel world of nature? Wasn't it my job to protect it? Much later, I realised that dealing with a loss such as this is part of being a beekeeper. You have to learn to accept that some hives survive and some die. My first hive taught me a lot about survival of the fittest.

Smaller losses are unavoidable, too. Even when visiting your hives for a quick check-up, you will accidentally squish a few bees. However, beekeepers cannot allow themselves to think about their bees as individuals: you must think about the hive as a whole. By making regular checks on your hives, you are protecting the other tens of thousands of bees – even if you do accidentally crush a few here and there. This was undoubtedly one of the hardest things for me to accept during my first few months of beekeeping. No one told me to think of my hives this way, but I eventually realised I had to.

Honey and Pollen

Contrary to popular belief, honey is not the only food that bees need. They also need pollen, which is their source of protein. Therefore, when checking in on your hives you should look for pollen stores in addition to honey.

Like honey, the bees store pollen in cells. It is quite easy to see due to its unique orange colour, rather than honey's characteristic golden yellow. Unlike honey, pollen is not always stored in a uniform fashion and will be scattered throughout the frames, but is still easy to find nonetheless.

Like humans, bees do not eat consistently. There are times of the year when they will eat more, and there are times when they will eat less. Whenever needed, I would recommend feeding your bees. However, knowing when to feed them is tricky. Often this comes with experience so, initially, work with a bee mentor who can help you identify what is happening in the hive and when you might need to intervene.

The simple syrup solution is a good source of food. Alternatively, if you lose a hive that had stored up honey before it collapsed, providing a struggling hive with a few frames of honey from it would be a beneficial move. Be careful, though: if the other hive died of a disease or some type of infestation, it would be unwise to move frames to another hive because it could simply transfer the infestation to your living hive. Lastly, it is also possible to feed them pollen. I use 'pollen patties' which you can make or buy.

Bee Gardens

In the summer months, bees are highly resourceful and will find food from nearby flowers. If, however, you are concerned there are not enough flowers nearby, you could consider planting a bee garden.

Planting bee-friendly flowers is a way of helping your local hives flourish and providing ongoing sources of nectar and pollen. When planting, be sure to keep in mind when each of the plants bloom so that you have a nice range of types and staggered bloom times. You want to ensure your bees have access to a good supply of pollen from spring until late summer. Plant native species, and opt for single flowers rather than those with multiple tightly-packed heads. For me that includes: foxglove, borage, lemon balm (commonly known as 'bee balm'!), zinnias, sedum, goldenrod, calendula, cosmos, echinacea, and hosta. We are lucky to have apple, avocado, orange and peach trees nearby as well. Also try Viper's bugloss *(Echium vulgare)* with its vivid blue flowers that will attract numerous bees in high summer, as well as comfrey, honeysuckle, and red and white clover. Avoid flowers that produce little or no pollen and nectar, such as pansies.

You could even plant little pots with various flowering plants – it doesn't have to be a big garden.

Do try to avoid using any pesticides in your garden, as almost all of these can harm bees. These products are often labelled as 'bug killers' or something similar.

Summary: Feeding Your Bees

— In general, bees are very self-sufficient in terms of feeding themselves.
— They produce and store honey to see them through the winter months.
— You may need to step in if you think your bees are struggling.
— A simple sugar solution usually works best.
— Plant a bee garden!

'His labor is a chant,
His idleness a tune;
Oh, for a bee's experience
Of clovers and of noon!'

—

Emily Dickinson

Let's have a look at what to expect from your first twelve months as a beekeeper.

Winter

As with most outdoor pursuits, beekeeping is heavily influenced by the four seasons. The winter is obviously the toughest and sadly some hives will die, but there are ways to get through even brutal winters in order to save your hives. However, these methods are not foolproof. This is a tough fact to swallow as a rookie beekeeper.

When I acquired my first three hives, I was convinced that I could prevent them from dying during the winter. I kept thinking to myself, 'It's just a temperature, how can I be beaten by a simple temperature?' The answer is this: as resilient as bees are, they simply cannot survive very cold winters. The winter of 2013–14 was especially brutal where I live in Newburyport, Massachusetts. Of my three hives that were living going into the winter, none survived. I took so many precautions. I shovelled out the front of each hive after every snowstorm in order to keep the moisture level down within each hive. I wrapped each hive with tarpaper to keep the temperature at a reasonable level. I placed Styrofoam covers under the inner cover, which provided insulation to the hives. Unfortunately, the winter was too long and too cold, and my efforts proved futile. While I knew what was coming, my heart sank every time I opened a hive and it was silent. I knew they were gone before I removed the cover, but the confirmation made me quite sad.

Spring

Fortunately, the spring and the summer are far more kind to your bees. During the spring, the time you spend on your bees will be mainly visiting them and checking the hive's health. You may have to make a few tweaks here and there, but nothing close to what you must do in preparation for the winter. If a hive has died over the winter, you must clean it out and salvage any useful parts that remain. Usually, you can reuse most of the frames and supers as a base for establishing a new hive. However, if your previous hive died due to a virus, bacteria or mite, it would be a sound idea to replace the parts when establishing a new hive.

Summer

During the summer, my main concern is ensuring that my hives are cool enough. This is why I recommend placing your hives near some trees or tall shrubbery to provide them with shade. If your hives overheat, the bees will have to expend a great deal of energy and time simply cooling the hive. This process can put extra strain on your bees if they begin working too hard to keep the hive cool – the overall health of the hive can suffer because of it.

Swarming

Swarming – when the queen leaves the hive with a majority of the worker bees and finds a new nesting site – is a way for a colony to manage growth and reproduce. It typically happens in spring.

The new site is often fairly nearby, in a tree perhaps, or a fallen log. The main criterion is that the new location can house enough bees and protect the swarm from the weather. Beekeepers try to prevent swarming because it leads to a very vulnerable time for a hive – the brood cycle is disrupted, and the remaining hive is usually left very depleted, which makes it quite unproductive with regards to honey.

There are a few ways for beekeepers to manage the hive in order to reduce the likelihood of a swarm. I think simple swarm management is part of the practice of beekeeping. The key thing to look out for is 'Queen cells' – these are cells where the hive is beginning to raise a new queen and can be a sign of imminent swarming. You can remove these as a preventative measure but you may not wish to interrupt the colony's natural activity. I'm of the opinion that a hive knows better than me what it needs. However, if the reason for swarming is overcrowding or poor ventilation, then you can be proactive. Always ensure you provide ample space for the brood so there is no overcrowding. The pressures of space can force bees to look for a new place to grow a hive. Also ensure you have adequate ventilation as poor air flow can be another reason for swarming. Finally, a hive that is about to swarm is a very healthy hive, with lots of activity.

Other than providing shade, the best way to prevent your hives overheating is to provide them with fresh, cool water. This means changing the water about once a week – assuming the only water they drink is that which you provide. This is why a hive near a lake or pond is ideal, because it has access to a constant source of fresh, clean water, which the bees use to cool the hive. All my hives are close to a body of water, yet I still supply the bees with water throughout the year. I would recommend you do this as well in order to keep your bees healthy, happy, and productive.

If you are supplying water, don't simply fill a bucket of water and place it in front of your hive. If you do this, the bees will drown as they attempt to drink and get caught in the surface tension of the water. During my freshman year of high school, a couple of hives were established on campus – with lots of help from many other students and faculty members. Fortunately, our school also has an outdoor swimming pool, which all the students use to cool off on hot days. However, this created an interesting problem. Since honeybees are attracted to chlorine, many of them casually buzzed their way over to the pool and dived right in. As other students were enjoying a nice swim, they began bumping into panicking honeybees. Realising what they'd just made contact with, they'd swim away as quickly as possible yelling, 'Orren, get rid of these bees!'

Autumn/Fall

At the end of the summer, right before autumn begins, you might notice a slight decrease in the population of your hive as the drones slowly die off. This is a natural process. The queen does not create more drones, because they consume the most honey on an individual basis by far.

The beginning of autumn is the perfect time to start making your winter preparations. If you are extracting honey, be sure to leave at least 75 to 100 pounds (35 to 45kg) for your bees to consume during the winter. This is extremely important. It could be easy for a young, inexperienced beekeeper to get carried away and accidentally remove too much honey. If anything, take less rather than more – there will always be next year!

The beginning of autumn is also when I might start to feed my bees my sugar syrup – the mixture of sugar and water (see Chapter 7). This mixture will provide enough energy for the bees to prevent them from prematurely consuming their honey stores. Keep a close eye on whatever mechanism you use to feed your bees. If the bees are taking the mixture quickly, there is a decent chance they are storing it away for later months, which can be dangerous. I have experienced hives doing this before, and for one, it ended up killing them.

The simple syrup does not have the same antibacterial and antiseptic properties as honey, meaning that mould and mildew can begin to grow within a hive that has too much sugar solution in storage. This also highlights the importance of proper ventilation within a hive, especially during the colder months when moisture levels naturally rise.

After addressing the potential issue of moisture within your hive, you might want to figure out a way to keep your bees warm before the cold winter months take hold. For beekeepers in warmer climates where it rarely drops below freezing, this next piece of advice is less applicable. For my hives in Massachusetts, I wrap tarpaper around the exterior of the hive before the weather becomes too cold. The black tarpaper serves as a slight windbreaker and, more importantly, it absorbs the heat from sunlight, therefore providing a little assistance to the bees as they attempt to keep their hive warm. This principle refers back to the location of your hives. Ideally, they will receive sun in the morning and shade in the afternoon. If this is achieved, the bees will get neither too cold during the winter nor too hot during the summer. If you live in an area with a warmer climate and moderate winters, prioritise giving your bees shade. Conversely, if the bees are likely to succumb to a long, cold winter, concentrate on providing them with more sun to aid them during this harsher season. A more extreme measure is to physically move your hives. I have never done this but commercial keepers frequently do – from the avocados in California to the blueberries in Maine, for example.

Summary: The First Year

Spring: A time for hive maintenance after the winter and reversing hive bodies.
Summer: Keep hives cool and ensure bees have access to water.
Autumn: Time to extract some honey! Be sure to leave enough to take your bees through the winter. Also consider feeding your bees a sugar solution.
Winter: Take steps to prepare your bees for cold weather by wrapping your hives.

9
**Troubleshooting:
Diseases and Pests**

> 'That which is not good for the beehive
> cannot be good for the bees.'
>
> ---
>
> Marcus Aurelius

As a beekeeper, one of your key tasks is to protect your bees from pests and diseases that could prove fatal to your hive.

American Foulbrood

This is the deadliest of all of the 'bee brood' diseases. As the name suggests, this disease attacks the brood and either kills them off or weakens them so significantly that they die soon after hatching. The bacteria attack the young brood, under 24 hours old, and slowly feed on the larva – changing it from a healthy white colour to a dying brown. The larvae die soon after the cell is capped, since the bacteria multiply very quickly. The bacteria spread from cell to cell because the nurse bees inadvertently transfer them during feeding. A hive infected with American foulbrood will appear rather patchy, since the dead brood are a darker colour than the healthy brood. Another telltale sign of the disease is the sunken appearance of the capped cells, unlike healthy capped brood, which protrude from the frame. Furthermore, the foulbrood will give off an unpleasant odour, which experienced beekeepers can smell.

European Foulbrood

This is another bacterial disease, and can be identified by a sour smell and ropey, stringy threads. It is important not to confuse European foulbrood with American foulbrood. Both are bacterial diseases but that is where the commonality ends. The first sign of EFB is often a spotty brood pattern. A simple test is to stick a toothpick into the cell and draw it out. If the remains are ropey and sour-smelling, this will indicate that something isn't quite right. The best way to absolutely confirm EFB is to have hives inspected by a qualified bee inspector.

Important

Both American and European foulbrood are notifiable diseases in England and Wales, and if you suspect you might have a case you must inform the National Bee Unit (www.nationalbeeunit.com).

Chalkbrood

Chalkbrood is another serious brood disease that can ruin a colony if not handled properly. Most hives will be affected by chalkbrood at some point during their lifetime; a stronger hive can simply push through the infection seemingly unaffected – as was the case with mine, thankfully. The chalkbrood fungus infects growing larvae when the spores attach themselves to the food given by the nurse bees to the larvae. As the larvae grow and develop, the spores increase in number, eventually killing them. The name chalkbrood originated from the white chalkiness of the larvae once they have died. It has the appearance of a little cotton ball at first. After some time, this changes into a dry, grey-coloured little Egyptian mummy. They really do look like little mummies, I've seen plenty of examples!

The spores responsible for this destruction are highly infectious. If equipment is taken from a previously infected hive and transferred to another, the disease will invariably take hold again. According to highly experienced beekeepers, the spores can even be transferred by foraging bees as they bring pollen from one plant to another. The spores latch on to the pollen, and if a bee from a neighbouring hive pollinates that same flower, the spores can be passed from one hive to another. The incredibly infectious nature of chalkbrood makes it a scary find for any beekeeper when opening a hive, especially if you have multiple hives all in the same area. As for managing chalkbrood, there are no medicines available of which I am currently aware, meaning that if a beekeeper finds chalkbrood in a hive, they must simply work their hardest to manage it by removing and replacing infected frames and being overly cautious when moving from one hive to another.

Sacbrood

Like the first two brood diseases, sacbrood will enter a cell before it is capped and attack the larvae. Once the larvae have died, the cell will fill with liquid. Sacbrood is a virus and can be spotted by the first symptom, which is dead or dying brood and larvae that have turned a grey colour as opposed to their usual bright white. The larvae skin becomes tough and the inside is watery and granular. The only way to treat sacbrood is to re-queen.

Mites

All beehives will have mites: some will just have worse cases than others. If you are uncertain about the status of mites within your hive, there is a straightforward way of finding out the severity of the problem. As the bees move and jostle around, mites will occasionally fall off and land on the bottom board. If you have a solid bottom board, it can be quite difficult to quantify the problem of mites within each hive. However, as you switch from a solid bottom board during the winter (which prevents draughts and helps to seal in the warmth) to a screened bottom board during the warmer months (which allows for ventilation and helps to cool the hive), you can use this as a way of assessing any mite problem. Many bee supply companies sell sticky white boards that have square-inch segments mapped out and numbered by row and column. These boards enable the beekeeper to get a general sense of how bad the mites are as they fall from the bees and onto the sticky bottom board. If there are more than 100 mites on the board in a 24-hour period, then you have a mite problem. If you are a brilliant note taker, you can even make a simple sketch or actually count the individual mites with a magnifying glass. In terms of treatment, a natural route is to have a screened bottom board which allows the mites to fall out of the hive.

Pests

One of the more entertaining yet unexpected problems that every beekeeper will encounter is pests. In this case, I am not talking about hornets, wasps, and other predators to the bees, but mice, ants, and other more benevolent animals who love your bees almost as much as you do.

Many times, I have opened a hive and found a dead mouse wrapped in propolis – which essentially mummifies the little body. Mice have been a prevalent problem for my hives, which are located at the edge of a field near woodland, the natural home to many mice. They seem to just wander up to the hive and sneak in. At first they are under the hive, then the little babies get inside. Unfortunately for the mice, the hive also happens to be full of bees who don't take kindly to unexpected visitors.

Pests I encounter more frequently – nearly every time I visit my hives – are ants. Ants often appear in a hive, attracted by the honey. They can build a little nest under the hive or in and around it. Ants don't seem to bother the bees and don't seem to cause much damage to the hive but they are certainly a nuisance.

Summary: Diseases and Pests

— Every beekeeper will encounter diseases and pests.
— Both American and European foulbrood are notifiable diseases in the UK.
— Regular inspection and maintenance can help you keep on top of most problems.

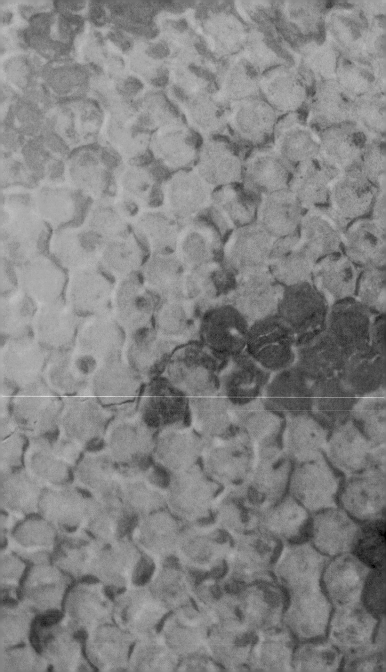

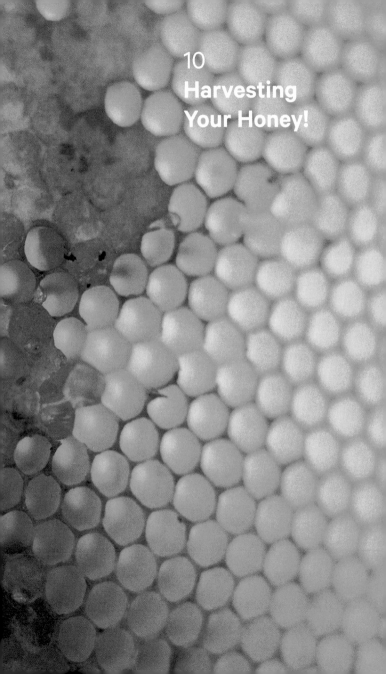

'You can thank a honeybee
for a third of your diet.'

———

Anon.

Now to the subject you have all been waiting for: Honey. Some beekeepers keep hives because they love the process of beekeeping, others do it for the pollination and the ecosystem, and some do it for the honey.

Whatever the motivation, the first honey harvest is one of the most satisfying moments any beekeeper will ever experience. I can clearly remember uncapping honey for the first time and seeing liquid gold spill out of a few cells. I slid my finger into its path and let the honey ooze over it. As I lifted my hand to my mouth and tasted honey from my hives for the first time, it tasted far better than any honey I had ever tasted before.

Honey History

Honey is a powerful substance; for centuries it has been used as both a sweetener and for its medicinal properties. The ancient Egyptians baked honey cakes that were offered to the gods to placate them. The Greeks used honey as a medicine; among other things, honey is said to be good for suppressing coughs, soothing burns, boosting immunity, and aiding those with seasonal allergies.

The thing that makes honey useful in the healing of wounds – its antimicrobial properties – also prevents the honey itself from going bad. As honey ages it crystallises, changing its physical properties while remaining perfectly fit to consume.

Honey comes in all different colours, viscosities, flavours, and scents. In some cases, it is possible to taste the different types of flowers that your bees have collected from when making their honey. For example, the flavour of orange blossom is quite strong in the honey from hives located near orange groves in California.

The honey from my hives in Massachusetts is said to taste like apples, peaches and the Atlantic Ocean. I love that description – even though the bees wouldn't be collecting anything from the ocean, it just sounds delicious. I usually harvest at the end of the summer months and, in a good year, I can harvest about 60 pounds (27kg) of honey from each hive.

How to Harvest

When harvesting your honey, there are many things to consider. First, you must decide weeks in advance whether or not a hive is fit to be harvested from. If a hive is well established and thriving it should be fine, but if it is a new hive or is struggling it might make sense to leave the honey this year to help your bees. Honey will mainly be harvested at the end of the summer or the beginning of the autumn. However, you must ensure that the hive has vast stores of honey, enough to last it through the winter. One of the worst things a beekeeper can do is to remove too much honey from a hive, depleting its resources and risking the hive dying from starvation.

If the hive is deemed fit for extraction, start by placing a honey super on top of the deep supers. The honey super is smaller than the deep super to enable the beekeeper to carry it when it's full of honey – it weighs in at roughly 50 to 60 pounds (23 to 27kg). A deep super full of honey would be too heavy to lift and, when spinning the honey out of the frame with an extractor, the frame might crack under the sheer weight and the force being exerted.

Some beekeepers place a queen excluder under the honey super, on top of the deep supers. This prevents the queen from climbing up into the honey super and laying eggs where only honey is desired. As we have discussed earlier, though, it can also slow down the speed at which the bees are able to work. In recent years, I have chosen not to use a queen excluder and, fortunately, I have experienced no ill effects. So far, I have yet to find any eggs in a honey super, and my bees have produced copious amounts of honey.

Once your bees have filled a honey super with honey, carefully lift it out and take it to a place where you can get really messy without ruining anything – a garage is perfect. Before you do this, however, you will need to overcome one obstacle: the bees. Months of tireless work from your bees have gone into making the honey you are about to harvest. It will be difficult to remove every single bee from the honey super before you begin to move it, but the process of harvesting the honey becomes increasingly simple when a cloud of bees isn't interrupting your work.

The method I use is to begin by smoking the hive before removing the honey super, as the smoke drives many of the bees away. Next, I brush the bees from each individual frame as I simultaneously shake them off over the hive. This entire process takes place directly over the hive for a few reasons. Firstly, if you choose not to use a queen excluder, the queen could be residing in the honey super.

If this is the case and you choose not to remove the bees over the hive, the queen could fall out from the super and land outside of the hive, which would be a HUGE problem. Secondly, by shaking the bees off back into their hive, you cause far fewer injuries while simultaneously helping them by relocating them back into the hive.

Extracting the Honey

For the extracting process, I'd advise that you rent a honey extractor for the summer. This is a device that spins uncapped frames of honey, forcing it out of each cell where it then accumulates in the bottom of the extractor.

There are two methods I use for uncapping a frame of honey before placing it in the extractor. One is to use an electric uncapping knife. The knife has a cord that is plugged into an outlet and provides heat to the blade. The heat allows the blade to slide through the wax easily, therefore uncapping the cells. The second method involves a tool called a capping scratcher. The scratcher has many small prongs on the end, which you use to scrape off the caps of each cell. The process is simple, but can be quite messy as the honey pours out.

Once the cells are uncapped, place each frame into its respective place in the centrifuge, the honey extractor. Now, the fun part. This step in the process is entertaining to watch as the person who is running the centrifuge works tirelessly to spin it as quickly as possible in order to force the honey out. This piece of the process does take some time, since honey is so thick, but results in an accumulation of honey in the bottom of the centrifuge. This honey will eventually be filtered in order to remove any debris that might have made it into the honey before it's poured into jars.

The final action is to open the valve at the bottom of the centrifuge and watch the honey ooze satisfyingly out. At this point in your beekeeping career, the sight of honey being poured into the first jar will be extremely gratifying. I remember this moment clearly: hours of hard work were finally paying off with a glorious jar of honey!

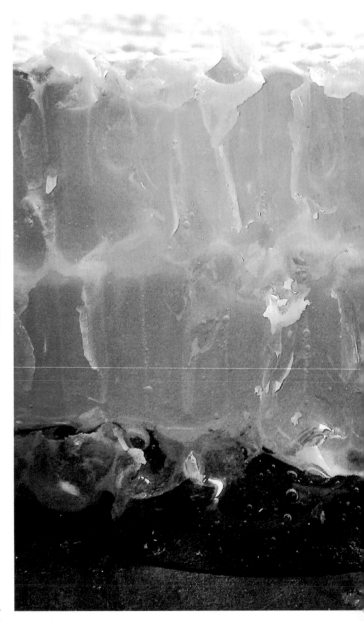

Types of Honey

There are several categories of honey: comb, liquid, set (or creamed), and chunk.

Comb honey is when the beekeeper harvests not only the honey but also a piece of the beeswax comb. Both the beeswax comb and the honey are edible!

Liquid honey is when, as described above, the beekeeper cuts the wax cappings off, the frame is placed into the extractor and honey is whizzed/spun out of the comb.

Set honey is when finely granulated honey is added to liquid honey and mixed until it is creamy. The benefit of set honey is that it is spreadable rather than pourable.

Lastly, **chunk honey** is a chunk of comb honey put into a jar with liquid honey poured all around it. This is a very, very beautiful way to bottle up your honey.

The most important thing to remember is to leave enough honey for your bees – about 75 to 100 pounds (35 to 45kg) – because at certain times of year it is your hives' only food source.

Summary: Harvesting Your Honey!

— Honey has been used both as a sweetener and for its medicinal properties for millennia.
— Be careful not to displace the queen when harvesting.
— Hire a honey extractor for the harvest.
— When removing honey from your hive, be sure to leave some for the bees!

' "Well," said Pooh, "what I like best," and then
he had to stop and think. Because although
Eating Honey was a very good thing to do,
there was a moment just before you began to
eat it which was better than when you were,
but he didn't know what it was called.'

A.A. Milne, *The House at Pooh Corner*

Recipes

Very Crunchy Honey Granola

This recipe is shared by Mollie Katzen, author of *Moosewood Cookbook* and *The Heart of the Plate*, who also said these kind words about my honey!

'With each golden drop, I can taste not only the sources of the nectar, but also the air, the Atlantic Ocean, and the devotion of this committed young beekeeper and farmer. Orren's BeeHappy Honey is a labor of love, and the flavor of joy.'

—

Ingredients

3 cups rolled oat flakes
1 cup barley flakes*
1 cup oat bran
1 cup sunflower seeds
1 cup chopped almonds
¾ cup canola oil
½ cup honey
1 tablespoon vanilla extract
1 cup soy protein powder
½ teaspoon salt
⅓ cup (packed) brown sugar
1 cup pumpkin seeds (optional, but highly recommended)
Non-stick spray for the baking tray

If you can't find barley flakes, you can substitute wheat flakes or just use 4 cups of rolled oats.

—

Method

1. Preheat oven to 325°F / 160°C / gas mark 3. Spray a 13×18-inch (33×45cm) baking tray with nonstick spray.

2. Combine the flakes, bran, sunflower seeds, and almonds in a large bowl.

3. Combine the oil, honey, and vanilla extract, and pour this in. Mix thoroughly. Use your hands if necessary.

4. Stir in the soy protein powder and salt, and mix thoroughly. Again, use your hands.

5. Place on the baking tray and bake for 35–45 minutes, or until golden. Stir it once or twice during the baking.

6. Crumble in sugar as soon as it comes out of the oven and let it melt in. Cool it on the tray, and stir in the pumpkin seeds as it cools. *Note: The granola will get crunchy as it cools.*

7. Store the finished granola in a tightly closed jar in the freezer for maximum freshness. This fits nicely into two 10oz (300g) jars.

Variation
Cool Berry Granola

You can add sliced fresh strawberries – or whole fresh raspberries or blueberries – directly to the granola before you freeze it. After the granola has cooled, add up to two cups of berries, stirring them in gently until the cereal surrounds them like a protective coating. Carefully pack the mixture into jars, close them tightly, and freeze. The berries will store beautifully this way, and will defrost very quickly in your cereal bowl, after you add milk.

Energy Bars

This recipe is shared by Sally Sampson, founder of ChopChopKids.

'These fruit-and-nut bars are super-easy to make (even kids can do it!) and easy to adapt to your own tastes. Love pecans? Use all pecans. More of a mixed-nut person? Use mixed nuts. Keep trying these bars with different ingredients: we also like toasted sesame seeds, sunflower seeds and/or pumpkin seeds in place of some of the nuts. We're not sure why, but every now and then these don't turn out quite right – they're crumbly instead of firm. If this happens, use it as a fruit or yogurt topping, and call it "granola"!'

—

Ingredients

½ cup lightly toasted nuts (one kind or a combination of almonds, walnuts, and pecans)

¾ cup dried fruit (one kind or a combination of raisins, currants, dried cranberries or chopped dates, prunes, apricots and peaches)

¾ cup quick-cooking oats

¾ cup crispy-rice cereal

2 tablespoons unsweetened coconut (if you like)

½ cup almond or peanut butter

¼ cup honey

½ teaspoon vanilla extract

—

Method

1. Line a baking tin with wax or parchment paper and leave enough hanging so that you can use it to cover the bars later (you will need a piece a little more than twice the size of the bottom of the tin).

2. To toast the nuts, put them on a small baking sheet in a 350°F / 180°C / gas mark 4 oven until they are fragrant and look a shade darker, around 5 minutes.

3. Put the nuts, dried fruit, oats, rice cereal and coconut in the bowl and toss well.

4. Put the almond or peanut butter and honey in the small bowl and microwave until the almond butter is softened, about 30 seconds (depending on your microwave). Stir until smooth. Add the vanilla and stir again until smooth.

5. Pour the almond mixture into a large bowl and mix with a large spoon until well combined.

6. Place the mixture into the prepared baking tin and pat down as hard as you can. You want to make the bars solid (rather than airy). Using the overhanging wax paper, cover the bars completely. Cover with clingfilm and refrigerate for at least 4 hours and up to one week.

7. Using the knife, cut into 16 pieces.

Honey Mustard Dressing

This recipe is one of my own.

—

Ingredients

5 tablespoons honey
3 tablespoons Dijon mustard
2 tablespoons vinegar

—

Method

Mix all ingredients together and whisk a little.
You can serve this as a dressing or a dip.

Famous Cooked Carrots

This is the first thing I ever cooked. I was four!

—

Ingredients

1lb (450g) baby carrots
3 tablespoons salted butter
3 tablespoons honey
1 tablespoon lemon juice
Pinch of crunch (coarse) salt

—

Method

1. In a medium saucepan bring some water to the boil, add the salt and then the baby carrots. Cook them for 4–5 minutes.

2. Drain the carrots and add to a cast-iron pan where the butter and honey are waiting. Cook until there is a nice glaze on the carrots.

3. Splash with the lemon juice then add a sprinkle of salt and pepper.

Honey Yogurt

This is a little obvious but it tastes awesome ...

—

Ingredients
A little bowl of Greek yogurt
3 tablespoons of local honey

—

Method

Mix it all about and feel free to add anything you like –
some nuts, raisins, flax seeds, granola, cut-up bananas,
berries or dried fruit. Delish!

Honey and Oat 'Chez Hay' Pancakes

This recipe is from my friend Hayley.

—

Ingredients
1 cup plain (all-purpose) flour
1 teaspoon baking powder
½ teaspoon baking soda
A small pinch of salt
1 cup buttermilk
¼ cup milk
2 tablespoons melted butter
1 tablespoon honey
1 large egg

—

Method

1. Mix the dry ingredients in a large bowl.

2. Mix the wet ingredients in a medium bowl.

3. Add the wet and dry together and stir vigorously until smooth.

4. Spoon the mixture onto a hot griddle and cook for about a minute or until golden brown on each side.

5. Drizzle aggressively with maple syrup – or honey!

Spiced Honey Cake

This recipe is from Bill Yosses, White House executive pastry chef. I met Mr Yosses while at an event at the White House called 'Know Your Farmer, Know Your Food', and he was kind enough to take us on a tour including the beehives out front.

—

Ingredients
3 cups plain (all-purpose) flour
½ teaspoon salt
1 teaspoon cinnamon
1 teaspoon baking soda
2 cups honey
2 eggs
1½ cups orange juice
½ cup raisins

—

Method

1. Preheat oven to 325°F / 160°C / gas mark 3. In a large bowl, sift together the flour, salt, cinnamon and baking soda, and set aside.

2. In another bowl mix the honey, eggs and orange juice with a wooden spoon until combined.

3. Add this mixture to the dry ingredients and mix together. Add the raisins.

4. Grease two 9×5-inch (22×13cm) loaf pans and divide the batter between the two. Bake for 75 minutes or until a toothpick inserted in the centre comes out clean.

Feta and Honey Cheesecake

This recipe is shared by the kind permission of
Sarit Packer and Itamar Srulovich, the founders of
Honey & Co., and can be found in their brilliant book
Honey & Co: Food from the Middle East (Saltyard Books).
It is one of their signature dishes.

—

Ingredients

For the kadaif pastry base
25g (1oz) melted butter
50g (2oz) kadaif pastry (or shredded filo pastry)
1 tablespoon caster sugar

For the cheesecake cream
160g (5oz) full fat cream cheese, like Philadelphia
160ml extra-thick double cream
40g (1½oz) icing sugar
40g (1½oz) honey of your choice (a grainy one works well)
50g (2oz) smooth, creamy feta
Seeds from ½ vanilla pod (or 1 teaspoon vanilla essence)

For the honey syrup
50ml honey
50ml water

For the garnish
A few fresh oregano or marjoram leaves
A handful of whole roasted almonds, roughly chopped
Some mellow-flavoured seasonal fruit – white peaches
    or blueberries are best (although raspberries or
    apricots are also good)

—

Method

1. Preheat the oven to 350°F / 180°C / gas mark 4.

2. Mix the melted butter with the pastry and sugar in a bowl. Fluff the pastry by pulling it and loosening the shreds with your hands till it gets an even coating of sugar and butter. Divide into four equal amounts, pulling each clump of pastry out of the mass like a little ball of yarn. Place these on a baking tray lined with parchment paper. They should resemble four flat birds' nests, each about the size of a drinks coaster.

3. Bake for about 12–15 minutes or until golden. Allow to cool and keep in an airtight container until ready to serve. The pastry nests will keep for 2–3 days, so you can prepare well in advance.

4. Place all the cheesecake cream ingredients in a large bowl and combine with a spatula or a big spoon, using circular folding motions until the mixture thickens and starts to hold the swirls. Don't use a whisk: it's vital not to add air to the mixture as the secret is in the texture. Check that it is sufficiently thick by scooping some onto a spoon and turning it upside-down: it should stay where it is. If it is still too soft, mix it some more. (If you are increasing the quantities in this recipe to feed lots of people, I suggest using a paddle on a mixer for this, but you'll need to watch it like a hawk so it doesn't turn into butter.) You can prepare the cheesecake cream in advance (up to 48 hours before serving) and keep it covered in the fridge until it is time to assemble the dessert.

5. Put the honey and water for the syrup in a small pan and boil together for 1 minute, skimming off any foam or impurities that come to the top. Remove from the hob and leave to cool, then store covered in the fridge until you are ready to serve.

6. When you come to assemble the dessert, place a pastry nest on each plate and top with a generous scoop of the cheesecake mix. Sprinkle over the herb leaves and chopped nuts, add a few blueberries or a couple of slices of peach, and drizzle a tablespoon of the honey syrup over everything. If you want to be super luxurious, drizzle with some raw honey as well.

Summer Honey Salad

This recipe is excerpted with kind permission from *The Honey Connoisseur: Selecting, Tasting, and Pairing Honey* by C. Marina Marchese and Kim Flottum (Black Dog & Leventhal).

—

Ingredients
2 large, ripe peaches
2 large tomatoes
¼ cup balsamic vinegar
2 teaspoons gallberry or star thistle honey
 (or your favourite type)
2 cloves garlic, chopped
½ cup olive oil
1 teaspoon Dijon mustard
Fresh basil, shredded

—

Method

1. Dice the peaches and tomatoes into bite-size pieces. Toss into a large bowl.

2. Put all the remaining ingredients for the dressing into a separate medium bowl and mix well.

3. Add the dressing to the tomatoes and peaches and toss until covered.

4. Serve at room temperature.

Fromageon

This recipe is from Elizabeth Gawthrop Riely, author of
A Feast of Fruits (Macmillan) and *The Chef's Companion*
(Wiley) and former editor of *The Culinary Times*.
She says:

*'Fromageon is a French term for goat cheese from the
Midi (south). In this version it is a fresh chèvre softened
with a little cream and flavored with honey. This rustic
dessert is at once very simple and quite sophisticated –
perfect for a relaxed evening of conversation among
friends. This fromageon is rich and aromatic, so serve
it in small portions or let your guests help themselves.'*

—

Ingredients
6oz (170g) fresh, mild goat cheese
About 4 tablespoons cream
1½ – 2 tablespoons honey (best quality)
Fresh fruit such as pears, apples, or grapes

—

Method

1. Put the chèvre in a small bowl and, using the back of
   a spoon, mash it with the cream to soften and blend
   the mixture.

2. Stir in the honey, holding some back so as not to over-
   sweeten. Taste carefully for balance, adding more
   honey if needed; remember that darker honeys are
   stronger in character.

3. Put the fromageon in a small crock or bowl and chill until serving time.

4. Serve the fromageon with fruit such as crisp pears and grapes, perhaps some toasted walnuts and a sprig of fresh rosemary for aroma and colour. Thin-sliced and toasted nut bread also makes a fine accompaniment, along with a glass of Armagnac or Madeira.

Green Beans and Radishes with Honey

Annie Novak, who kindly shared this recipe, is the founder and director of Growing Chefs, a field-to-fork food education programme, and the manager of the Edible Academy at the New York Botanical Garden. Annie writes in her own words for the *Atlantic*:

'Years ago, while working with farmers in New Zealand, I was told by a chef that the secret to a good recipe is to cover all your bases: hit every taste – sweet, salty, savory, spicy and sour – and keep an eye on cooking for satisfying mouth feel, or texture. This Asian-influenced recipe answers to all of that, and you'll look like a very sophisticated cook indeed when you get doubters to fall in love with the humble radish. As summer heat turns your radishes spicy in the soil, you need to borrow the sweet-with-a-kick cloak of chili honey to cover up their sharp bite.'

—

Ingredients

1lbs (450g) green beans, trimmed
¼ cup extra-virgin olive oil
8oz (225g) radishes, trimmed and quartered
2 cloves garlic, chopped
1 tablespoon honey
1 teaspoon chili flakes
Salt and freshly ground black pepper, to taste

—

Method

1. Bring a large pot of water to the boil over a high heat. Blanch green beans until crisp-tender, 3–4 minutes.

2. Remove from the heat and plunge the green beans into an ice bath to stop them from cooking further. When cool enough to touch, cut into bite-sized pieces if desired (you can also serve them whole, trimmed).

3. Heat the oil in a large frying pan over a medium heat. Add garlic and cook until just golden, about 2 minutes.

4. Add the green beans and radishes, and cook until the vegetables are easily pierced with a fork, about 5 minutes.

5. Add honey and chili flakes, continually stirring so the honey does not burn. Add salt and pepper to taste, and cook until the vegetables are just beginning to caramelise, about 2–3 minutes more.

6. Transfer the salad to a large bowl; set aside to let it cool slightly.

7. This can be served warm or at room temperature.

Cauliflower with Grapes and Honey

Another fine recipe shared by Annie Novak. This one combines seasonal crops brought together during a harvest: cauliflower and sweet late-summer grapes.

'This recipe draws from many childhood memories of Latin class, obsessing over the decadent meals served at the tables of mortal and mythological figures alike. Cicero and Pliny both praised honey, and Virgil, too, most hyperbolically and aptly, "Next I sing of honey, that heavenly, ethereal gift".'

—

Ingredients

1 head cauliflower
1 bunch green grapes
2 cloves garlic
½ cup pine nuts (pistachio nuts or walnuts are also delicious)
3 tablespoons olive oil
1–2 tablespoons fennel seed, to taste
1–2 tablespoons honey, to taste
Salt, to taste

—

Method

1. Preheat the oven to 375°F / 190°C / gas mark 5.

2. Wash and chop the cauliflower into bite-sized pieces. Wash the grapes and cut in half. Mince garlic. Roughly chop nuts.

3. In a large bowl, toss the cauliflower, grapes, garlic, nuts, fennel seeds, and olive oil. Salt to taste.

4. Pour the mix into a casserole dish, cover with tinfoil, and bake for 20–25 minutes or until the cauliflower is soft.

5. Remove tinfoil and bake for an additional 10 minutes, or until nicely browned.

6. Add honey and salt; toss and serve.

Moroccan-Style Winter Vegetable Stew

Ramin Ganeshram is a chef and food writer. Her most recent book *Future Chefs* (Rodale) highlights the achievements of young cooks, activists and reformers across America and the world.

'This stew is inspired by the warmly spiced and honey-sweetened tagines of Morocco. Hardy winter vegetables work best and you may substitute or experiment with your favourites. Winter pumpkins, Brussel sprouts, and sweet potatoes are just some examples of vegetables that would work equally well in this dish.'

—

Ingredients

1 cup dried chickpeas or one 15oz (400g) can chickpeas
½ teaspoon baking soda
1 tablespoon extra-virgin olive oil or, preferably, food-grade argan oil
1 small onion, sliced
1 tablespoon freshly grated ginger
1 parsnip, trimmed and chopped into 1-inch (2.5cm) pieces
2 carrots, trimmed and chopped into 1-inch (2.5cm) pieces
1 small head cauliflower, florets trimmed into 1-inch (2.5cm) pieces
1 large courgette (zucchini), chopped into 1-inch (2.5cm) pieces
1 teaspoon turmeric
½ teaspoon chili powder
½ teaspoon cinnamon
½ teaspoon cumin
1 teaspoon freshly ground black pepper
1 cup stewed, roasted tomatoes, diced

1 small anise pod
2 whole cloves
2 cardamom pods, slightly crushed
2 cups vegetable stock
1 teaspoon coarse salt or to taste

—

Method

1. Prepare the dried chickpeas: soak the chickpeas in 3 cups of cold water overnight. Drain the water and bring 3 cups of water to the boil in a medium saucepan. Add the drained chickpeas and the baking soda and lower heat to a simmer. Simmer for 30 minutes or until the chickpeas are tender. Drain and set aside. Alternatively, you may use canned chickpeas: rinse the chickpeas in a strainer and set aside.

2. Preheat the oven to 350°F / 180°C / gas mark 4. Heat a Dutch oven or a large, deep, oven-safe saucepan over medium heat and add the oil. Add the onion and ginger and sauté for 2–3 minutes, or until the onion begins to soften.

3. Add the parsnip, carrots and cauliflower florets. Stir well and cook for 4–5 minutes or until the carrots begin to brown.

4. Add the turmeric, chili powder, cinnamon, cumin and black pepper and stir well. Cook the stew for about 1 minute or until the spices begin to release their aromas.

5. Add the tomatoes and stir well. Add the cloves and cardamom pods and stir well.

6. Add the vegetable stock, honey and chickpeas and

stir well. Simmer the stew for 10 minutes uncovered, then cover and place in the oven for 20 minutes more, or until the liquid is thickened and reduced by one-third. Add the salt and cover. Cook for 10 minutes more.

Serve hot with couscous prepared as follows:

—

Ingredients
1 teaspoon extra-virgin olive oil, or preferably argan oil
2 garlic cloves, minced finely
1½ cups vegetable stock
1 small bay leaf
¼ teaspoon turmeric
¼ teaspoon salt
¼ teaspoon freshly ground black pepper
1 cup whole wheat couscous

—

Method

1. Heat a small saucepan over a medium heat and add the oil. Add the garlic and sauté for 1 or 2 minutes.

2. Add the vegetable stock, bay leaf, turmeric, salt and black pepper. Simmer for 10 minutes.

3. Add the couscous and cook for 1 minute. Turn off the heat and place the lid tightly on the saucepan. Allow to steep for 10–15 minutes or until all the liquid is absorbed.

4. Remove the lid and fluff the couscous with a fork. Serve with the vegetable stew.

Comfrey Healing Salve

From my bee mentors Jane and Rob Wild, a non-edible recipe that is invaluable for chapped hands, dry lips and general loveliness. Jane is a 'backyard beekeeper', and vice-president of the Essex County Beekeepers' Association in Massachusetts. She has been keeping bees since 1991, and she and husband Rob have fourteen hives. I count on both of them every bee season and have probably asked them more questions than they thought possible.

—

Ingredients

6 teaspoons ground comfrey root
4oz (115g) beeswax
4oz (115g) olive oil
10 drops vitamin E oil

—

Method

1. Mix the comfrey with the olive oil. Add the beeswax and melt together in a double-boiler over a low heat, stirring well.

2. Remove from the heat, add the vitamin E oil and mix well.

3. Strain through a fine-mesh polyester or nylon bag (comfrey tends to clump in the wax, and the mixture may turn an off-grey colour).

4. When cool, store in a tightly sealed tin or wide-mouth jar (less beeswax will make the mix softer).

Resources

Online

BeeSource online forum
www.beesource.com
This website was started in 1997 by a hobbyist beekeeper; it became an online community for beekeepers and beekeeping in 1999. Today it has over 14,000 registered members and is the most active online beekeeping community of its kind in the world.

British Beekeepers Association (BBKA)
www.bbka.org.uk
Set up in 1874, this is the UK's leading organisation representing beekeepers.

Getting started – equipment
www.ent.uga.edu/Bees/get-started/equipment.html

Storey's Guide to Keeping Honey Bees
apis.shorturl.com

Dances with Bees
www.pbs.org/wgbh/nova/bees/dances.html

Anatomy of a Hive
www.pbs.org/wgbh/nova/bees/hive.html

Bad Beekeeping
www.badbeekeeping.com/weblinks.htm
This site provides a very comprehensive list of links.

Books

The Beekeeper's Handbook, Diana Sammataro, Alphonse Avitabile, Dewey Caron (4th edition, Comstock, 2011)

The Backyard Beekeeper, Kim Flottum (revised and updated, Quarry Books, 2010)

The Beekeeper's Bible: Bees, Honey, Recipes and Other Home Uses, Richard Jones and Sharon Sweeney-Lynch (Stewart, Tabori and Chang, 2011)

A Book of Bees: And How to Keep Them, Sue Hubbell (Mariner Books, 1998)

The Beekeeper's Lament: How One Man and Half a Billion Honey Bees Help Feed America, Hannah Nordhaus (Harper Perennial, 2011)

Honeybee: Lessons from an Accidental Beekeeper, C. Marina Marchese (Black Dog & Leventhal, 2011)

Better Beekeeping, Kim Flottum (Quarry Books, 2011)

Videos/Movies

Marla Spivak: Why bees are disappearing
www.ted.com/talks/marla_spivak_why_bees_are_disappearing

Backwards Beekeepers: How To videos
www.backwardsbeekeepers.com

More than Honey
buy.morethanhoneyfilm.com

Vanishing of the Bees
www.vanishingbees.com

Queen of the Sun
www.queenofthesun.com

About the author

Orren Fox is an 18-year-old beekeeper, chicken farmer, sustainable-food advocate, longboard builder and student. He grew up in Newburyport, MA, and has been keeping chickens and bees for most of his life. He was a speaker at the Do Lectures USA in 2012. Orren is the author of happychickens.com, a blog on raising healthy chickens and bees, naturally. He has written several articles for *Edible Boston*, Civileats.com and *Handpicked Nation* and has been interviewed for *The Huffington Post*, *Yankee* magazine and *Boston Globe*.

In May 2012, Orren was invited to the White House as the guest of Kathleen Merrigan, then Deputy Secretary of Agriculture, as part of a 'Know Your Farmer, Know Your Food' event. As part of the event, Orren was invited to send his honey to Sam Kass, the executive director of Let's Move and Senior Policy Advisor for Nutrition Policy, for a 'honey showdown'. The White House has hives and harvests the honey each year.

Orren is also the founder of #beechat, a Twitter meet-up whose goal is to gather beekeepers from around the world to share information and ultimately help bees. He is also launching an app called Beehaviors which will harness the power of citizen scientists to provide ongoing data to entomologists to help understand what is happening to the bee population around the US.

You can connect with Orren on Twitter: *@happyhoneybees* *@happychickens* or via email: *thehappychickens@gmail.com*

Surkhet, Nepal. Photo © Blink Now

Thanks

At the Do Lectures and Do Books, thanks to David and Clare Hieatt, Duke Stump, Anna Beuselinck and Miranda West for giving me the opportunity to be surrounded by such inspiring people; Michael Piazza (photographer extraordinaire); Jen Reddy (for helping me find a story to tell) and Maggie Doyne for allowing me to think anything is possible, including beekeeping in Nepal!

At school, thanks to: Mr Robinson, my amazing English teacher, who has taught me to be a more logical writer; Tim Sullivan, my advisor at school, who has really encouraged me; the Independent Committee at school for allowing me the time while at school to work on this book (I know you initially thought I was kidding); Mr Okin; Mr Doyle; Mr Meyer; and the Environmental Action Committee for starting up a beekeeping practice at the Thatcher School.

For recipes, thanks to Sally Sampson, Mollie Katzen, Bill Yosses, Marina Marchese, Hayley Willner, Ilene Bezahler, Annie Novak, Kim Flottum, Elizabeth Gawthrop Riely, Ramin Ganeshram, Sarit Packer and Itamar Srulovich.

Thanks to my 'mentors', Jane and Rob Wild from Essex County Beekeepers, and Ilene Bezahler from *Edible Boston*. To my friends, Dorothy Fairweather, Lisa Buczinski, Julie, and to David and Lisa Hall for allowing me to 'YardShare'.

And of course, my family. Alice Delana, otherwise known as Grandhoney, my brother, and my mom and dad!

Index

Books in the series:

Available in print and digital formats
from bookshops, online retailers
or via our website:
thedobook.co

To hear about events and
forthcoming titles, you can find us on
Twitter @dobookco, Facebook
or subscribe to our online newsletter.